NIST Technical Note 1700

# Performance of RFID Tags in Rough Duty Environments (Structural Fires and Moisture)

Jonathan Kent
J. Randall Lawson
Anthony D. Putorti
*Fire Research Division*
*Engineering Laboratory*

May 2011

**U.S. Department of Commerce**
*Gary Locke, Secretary*
National Institute of Standards and Technology
*Patrick Gallagher, Director*

# Performance of RFID Tags in Rough Duty Environments (Structural Fires and Moisture)

By

Jonathan Kent
J. Randall Lawson
Anthony D. Putorti

## *Abstract*

Radio frequency identification (RFID) tags have become widely used by industry, retail sales businesses, and government agencies for tracking materials, products, and inventories. This tracking technology is starting to be applied to the management of emergency responder protective equipment items. This technology has potential for helping to manage the use of emergency apparatus and may become a component of human body worn tracking and locating systems. Current standards governing the application of RFID technology are focused on the industrial sector, and no standards exist for use of the technology in the emergency response community. The National Institute of Standards and Technology (NIST) has conducted a series of five large-scale fire tests to measure the performance of RFID tags in elevated temperature environments that may be produced by structural fires. Passive and active RFID tag designs were evaluated. Data were gathered from each of these large-scale fire tests. Additionally, small scale experiments were conducted to better understand the response of these devices when exposed to elevated temperature environments. Small scale tests were also conducted to develop a basic understanding of RFID tag performance when exposed to conditions representative of wet personal protective clothing. Results from theses evaluations show that RFID tags are sensitive to elevated temperatures, and they can be destroyed if directly exposed to room fire environments. However, results also show that RFID tags may still function if they are protected by insulating materials. Experiments with wet clothing showed that passive RFID tags would not transmit more than a few millimeters when located in a wet garment. Active RFID tags continued to work while contained in wet clothing with a small loss in communications range. These experiments have shown some of the limitations of RFID tags under potential rough duty environments involving moisture and heat. Signal attenuation due to wet material precludes the passive tags being worn under PPE due to the potential of that fabric to be wet. The thermal failure level of RFID tags is low compared to other equipment standards, such as those for Personal Alert Safety Systems (PASS). As currently designed and protected, the RFID tags would not be able to operate at a temperature of 260 °C (500 °F) for five minutes, and therefore could not be used as a life safety device for a structural fire fighter. Further research is needed to develop means of protecting these devices from thermal hazard.

**KEY WORDS:** Emergency responder, fire fighter, identification, radio frequency, safety, standards, fire test

# Disclaimer

Certain trade names or company products are mentioned in the text to specify adequately the experimental procedure and equipment used. In no case does such identification imply recommendation or endorsement by the National Institute of Standards and Technology, nor does it imply that the equipment is the best available for the purpose.

Regarding Non-Metric Units: The policy of the National Institute of Standards and Technology is to use metric units in all its published materials. To aid the understanding of this report, in most cases, measurements are reported in both metric and U.S. customary units.

# Table of Contents

Disclaimer ............................................................................................................................. iii
Table of Contents ................................................................................................................ iv
List of Figures ...................................................................................................................... v
1 Introduction ..................................................................................................................... 1
2 RFID Tag Thermal Testing ............................................................................................. 3
   2.1 Background for Bench-Scale Thermal Testing ......................................................... 3
   2.2 Radiant Panel Test Procedure ................................................................................... 4
   2.3 Radiant Panel Test Results ........................................................................................ 5
3 Large-Scale Fire Tests ..................................................................................................... 5
   3.1 Test Specimens .......................................................................................................... 5
   3.2 Test Instrumentation ................................................................................................. 8
      3.2.1 Passive Tag Instrumentation ............................................................................. 8
      3.2.2 Active Tag Instrumentation .............................................................................. 9
      3.2.3 Measurement Uncertainty ................................................................................ 9
   3.3 Descriptions of Large-Scale Fire Tests ................................................................... 11
      3.3.1 Initial Fire Tests #1 and #2 ............................................................................. 14
      3.3.2 Fire Test #3 ..................................................................................................... 14
      3.3.3 Fire Test #4 ..................................................................................................... 15
      3.3.4 Fire Test #5 ..................................................................................................... 15
   3.4 Test Results and Discussion .................................................................................... 15
      3.4.1 Results of Initial Fire Tests #1 and #2 ............................................................ 15
      3.4.2 Results of Final Large-scale Fire Tests .......................................................... 16
4 RFID Tag, Wet Experiments ......................................................................................... 22
   4.1 Passive Tag Systems ............................................................................................... 22
      4.1.1 RFID Equipment Used ................................................................................... 22
      4.1.2 Experiments .................................................................................................... 23
   4.2 Active Tag Systems ................................................................................................. 25
      4.2.1 RFID Equipment Used ................................................................................... 25
      4.2.2 Experiments .................................................................................................... 26
   4.3 Summary of Wet Experiment Results .................................................................... 26
5 Discussion and Conclusions .......................................................................................... 27
   5.1 RFID Tags and Emergency Response Operations ................................................. 28
   5.2 Implications for Body Worn RFID Tag Systems ................................................... 29
   5.3 Recommendations for Future Work ........................................................................ 30
6 Acknowledgements ....................................................................................................... 31

# List of Figures

Figure 1-1. Photographs showing examples of passive tags (left) and active tags (right) with no instrumentation. ................................................................. 3
Figure 2-1. Photographs showing thermal test apparatus with calibration assembly (left); RFID tag and thermocouple attached to substrate during a test (right). ...... 5
Figure 2-2. Photographs show examples of Passive RFID tags before fire testing. .......... 7
Figure 2-3. Active (top) and Passive (center and bottom) RFID tags before fire testing. .. 7
Figure 2-4. Photographs showing passive antenna (top left) active antenna and receiver (top right) and computers used for data logging (bottom). ....................... 11
Figure 2-5. Diagram of fire test facility layout and final location for testing RFID tags. The diagram shows the bedroom (top), living room (bottom), and center hallway connecting them. Note: Sketch is not to scale. ........................... 12
Figure 2-6. Photographs show the bedroom (top left), the living room (top right), the center hallway (center left) before the test; the fire in the bedroom (center right), the bedroom (lower left) and the end corridor (lower right) after the fire test. Tags PA 1 and PV 1 are shown in Figure 2-2 and Figure 2-3... 13
Figure 2-7. Photographs showing passive RFID tag after fire exposure, the tags on the left specimen board had burned away. ................................................. 16
Figure 2-8. Test 3, temperatures center of south corridor housing RFID tags. ................ 17
Figure 2-9. Test 3, temperatures on wall next to RFID tag specimens. Loss of signal (LOS) was observed for passive tags; LOS was not observed for active tags. ................................................................................................. 18
Figure 2-10. Test 4, temperatures center of south corridor housing RFID tags. ............. 18
Figure 2-11. Test 4, temperatures on wall next to RFID tag specimens. Loss of signal (LOS) was observed for passive tags. ................................................... 19
Figure 2-12. Test 5, temperatures center of south corridor housing RFID tags. ............. 19
Figure 2-13. Test 5, temperatures on wall next to RFID tag specimens. No passive tags were tested; loss of signal (LOS) was observed for the active tags. .......... 20
Figure 2-14. Photographs showing fire test damage to passive and active RFID tags. Note: RFID tag identification for these photos is found in Figure 2-2 and Figure 2-3. ............................................................................................. 22
Figure 3-1. Passive RFID tag reader antenna used for wet fabric study ......................... 23
Figure 3-2. Photographs showing passive (PA1) RFID tags attached to cotton T-shirt fabric (left) and fabric folded over tags (right). ...................................... 24
Figure 3-3. RFID tag (PV1) encapsulated in plastic with T-shirt fabric (left); RFID tag enclosed in dry T-shirt fabric (right). .................................................... 25
Figure 3-4. Active RFID tag (AW1) with T-shirt fabric (left) and RFID tag enclosed in wet T-shirt fabric (right). ...................................................................... 25
Figure 3-5. Photograph showing the complete active RFID reader system and one of the wet test specimens. ............................................................................... 26

# 1 Introduction

Radio frequency identification (RFID) is a technology that provides a means for using a computer to program identification and other information into a tag that can retain the data and wirelessly transmit the information back to a compatible electronic reader. RFID tags have gained wide usage throughout the retail sales market and are being used for tracking materials being shipped throughout North American and the world. Corporations that maintain large stocks of inventory are using RFID tags for inventory control, and the United States Department of Defense has begun using RFID as a tool for managing materials logistics.[1,2] Standards for RFID technology are presently focused on use by industry and the logistical chain associated with product inventory and sales. However, RFID technology is being recognized by the emergency response community as potentially useful for managing incident operations and assisting in locating equipment and emergency responders. In fireground situations, body worn RFID may be a tool that will improve fire fighter safety by helping to track and locate fire fighters while they are actively conducting rescue and fire fighting operations. RFID operational technology is advancing at a rapid pace with new prototypes and products being added to the marketplace annually. Hybrid RFID tag systems are now being produced that combine with other technologies such as environmental sensors, Global Positioning Systems (GPS), and cellular communications systems for providing tracking and situational information.[3] However, little is known about the performance of RFID tag technology when exposed to thermal environments and wet conditions that may be experienced by fire fighters and other emergency responders. These systems must be resilient when operating in these challenging emergency response conditions.

Currently, there are two basic types of RFID tags: "Passive" and "Active." For "Passive Tags" the RFID industry is presently marketing devices that are referred to as Gen 2 Tags. These tags represent a second generation of passive RFID design technology. The different types of RFID tags typically operate on different radio frequencies. Most of the tags presently being used throughout North America operate on the FCC (Federal Communications Commission) assigned UHF (ultra high frequency) band either in the 400 MHz or the 900 MHz frequency ranges. Additionally, a new form of RFID tag that uses ultra-wideband radio communications is under development. This tag, referred to as the Utag, is being developed to overcome some of the inherent shortcomings of the current RFID tag systems that are in wide use.[4] The Utags are currently being developed for use by the Department of Defense and various other government agencies and may

---

[1] Mount, David J., "RFID Tags – They're Everywhere!!!" Vacuum Technology & Coating, September 2007.
[2] Erwin, Sandra I., "Defense Dept. Begins New Effort to Better Track Military Supplies," *National Defense*, National Defense Industrial Association, Arlington, VA, September 2007.
[3] Bacheldor, Beth, "Hybrid Tag Includes Active RFID, GPS, Satellite and Sensors," *RFID Journal*, www.rfidjournal.com, February 2009.
[4] Nekoogar, Faranak, "Wireless That Works," Research Highlights, Science and Technology Review, Lawrence Livermore National Laboratory, December 2007.

not be immediately available to the emergency responder community. All of the tag types used in these experiments are listed in Table 1-1.

A passive tag consists of a small antenna connected to a small integrated circuit. These systems may be as thin as a few sheets of paper or several millimeters thick depending the design. The thin tags are often applied to a plastic film and have a sticky side that is used to attach to a material. The thicker tags are generally special purpose tags for attachment to metal containers or are designed to be high temperature tags. These devices typically measure a few millimeters wide and are often several tens of millimeters long. Figure 1-1 shows passive RFID tags in the left photograph with a shipping container type tag at the top and the more conventional plastic film tags below. A 15 cm long and 6 inch long ruler and a United States quarter coin are located in the pictures for comparison. The systems operate by absorbing energy from a RFID reader that transmits a radio signal to the tag. A tag receives energy from the reader signal, quickly gains enough power to transmit its programmed information back to the reader, and then transmits the data to the reader that is interrogating the tag. The most common passive tags typically operate in the 900 MHz radio frequency range.

An active tag consists of a small antenna connected to a small integrated circuit that is powered by a battery. As with the passive tag, the active tag is programmed with identification data. Power from the battery is used to produce the transmission signal and the signal is periodically produced at a set rate. Active RFID tags are generally larger than passive tags because they contain a battery. The right photograph in Figure 1-1 shows two different active RFID tag designs. The active tags often measure several tens of millimeters in width and length and are usually several millimeters thick.

**Table 1-1. RFID tags used in these experiments and the abbreviations used to identify them throughout this report. The estimated expanded uncertainty of distance measurements in this table is 1.0 %.**

| Abbr. | Type | Model | Dimensions (mm) | | |
|---|---|---|---|---|---|
| PA1 | Passive | Alien Technology ALL-9440 | 102 | 13 | Thin |
| PA2 | Passive | Alien Technology ALL-9460 | 76 | 76 | Thin |
| PA3 | Passive | Alien Technology ALL-9354 | 94 | 24 | Thin |
| PA4 | Passive | Alien Technology ALL-9338 | 98 | 30 | Thin |
| PI1 | Passive | Intermec Technologies Corp. Rigid Large | 32 | 156 | 10 |
| PD1 | Passive | Dynasys, Texas Instruments, High Temp. | 19 | 89 | 3 |
| PS1 | Passive | Symbol , type SYM 4x4 , Dual Dipole | 95 | 95 | Thin |
| PTI | Passive | Texas Instruments, Flexible Film Dipole | 38 | 95 | Thin |
| PV1 | Passive | Vanguard, Plastic Encapsulated, NIST logo | 102 | 102 | Thin |
| AW1 | Active | Wavetrend Personal identification tag | 29 | 64 | 10 |
| AW2 | Active | Wavetrend Mobile, Wire Antenna | 54 | 86 | 6 |

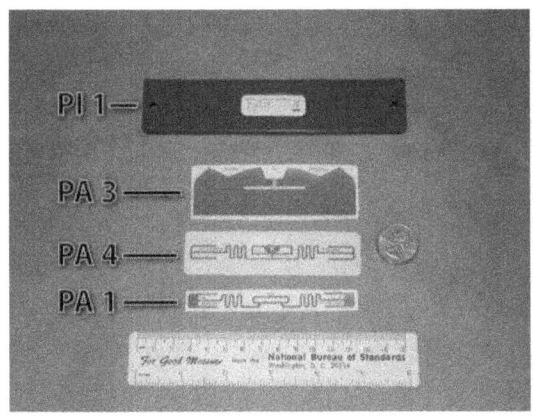 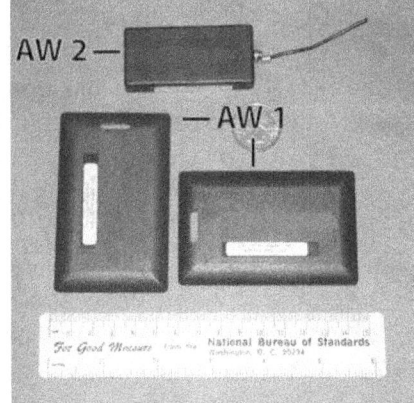

Figure 1-1. Photographs showing examples of passive tags (left) and active tags (right) with no instrumentation.

# 2 RFID Tag Thermal Testing

RFID tag systems that may be used by the emergency response community need to be resilient and able to function in a range of adverse environmental conditions. For the fire service, there needs to be an understanding of how well RFID tag systems perform in elevated temperature conditions that may be associated with fires. NIST has carried out a test program that involves RFID tags being exposed to bench-scale and large-scale thermal exposure conditions.

## 2.1 Background for Bench-Scale Thermal Testing

Small scale laboratory thermal exposure tests were carried out to obtain some baseline data on the performance of passive RFID tag systems prior to testing the RFID tag systems in large-scale fire tests. Several different test apparatus were available for providing the challenging thermal environments for testing the thermal response of RFID tag systems. One was a closed thermal chamber, and the second was an open test apparatus, both provided elevated temperature environments.

An example of the closed system is the "Thermal Flow Loop Test Apparatus," described by Donnelly et. al.[5] This test apparatus consists of a closed test chamber that uses a fan system to circulate heated air over a test specimen. This chamber is constructed of metal with a high temperature glass viewing window with minimal openings. The apparatus has proved to be effective when testing numerous types of electronic devices used by emergency responders.[5,6]

---

[5] Donnelly, M.K.; Davis, W.D.; Lawson, J. R.; Selepak, M.; "Thermal Environment for Electronic Equipment Used by First Responders," NIST TN 1474, 41 p., January 2006.
[6] Davis, W. D.; Donnelly, M. K.; Selepak, M.; "Testing of Portable Radios in a Fire Fighting Environment," NIST TN 1477, National Institute of Standards and Technology, Gaithersburg, MD, August 2006.

The second type of thermal test apparatus is an open system that uses a gas fired radiant panel to provide a thermal flux that can cause a test specimen to experience a controlled temperature increase. This apparatus is described by Lawson and Twilley.[7] This apparatus does not contain an enclosure and allows free access to materials being tested, as shown in Figure 2-1. The test apparatus has been effective when testing the thermal performance of a wide range of emergency responder protective equipment.

These apparatuses were used to conduct preliminary tests on the thermal performance of RFID tag systems. Additionally, the decision process on the selection of the correct test apparatus had to include the normal operations and performance characteristics of various RFID tag systems. These systems generally operate using RF transmit power levels that are considered to be low, usually at levels significantly less than one watt. Also, RFID tag systems can be affected by RF reflective materials and RF absorbing materials, and normally the RFID tags can be affected by their locations and the directional properties of the antenna systems. The open test apparatus more closely replicates the open test environment that was used in the large-scale fire tests. For these reasons, the preliminary thermal tests were conducted in the NIST Thermal Properties and Surface Flammability laboratory using a gas fired radiant panel system.[7]

## 2.2 Radiant Panel Test Procedure

Figure 2-1 shows the radiant panel with a calibration board and heat flux gauge located on the test frame. Figure 2-1 also shows the test setup with the RFID reader placed just to the left of the substrate board that has the RFID tag and a thermocouple taped to the board's surface next to the RFID tag. The types of tags tested and the number of tests are listed in Table 2-1. Note that the RFID tag is located on the side opposite from the radiant panel allowing the tag to heat from the substrate's surface. A total heat flux of approximately 2 $kW/m^2$, on the radiant panel side of the substrate board, was used for heating the test specimens. An aluminum radiation shield protected the substrate, tag, and thermocouple from heating until the measurements were to be made. At that point in time, the radiation shield was removed allowing the substrate material to heat. The Type K, 0.254 mm (0.010 in) nominal diameter wire, thermocouple attached to the substrate material had a digital thermometer attached to the wire and temperatures were observed and recorded.

---

[7] Lawson, J. Randall, and Twilley, William H., "Development of an Apparatus for Measuring the Thermal Performance of Fire Fighters' Protective Clothing," NISTIR 6400, National Institute of Standards and Technology, Gaithersburg, MD, October 1999.

Figure 2-1. Photographs showing thermal test apparatus with calibration assembly (left); RFID tag and thermocouple attached to substrate during a test (right).

## 2.3 Radiant Panel Test Results

Data from these tests showed that the signal strength and quality of Gen 2 passive RFID tags deteriorated such that the reader used could no longer detect it at temperatures of approximately 60 °C (140 °F). However, most of the RFID tags that failed at elevated temperature exposure started functioning again after cooling below 60 °C (140 °F). This information provided the basic starting point for evaluations to be conducted with the large-scale fire tests. The expanded uncertainty, $U$, for temperature measurements in this section are estimated as 6 °C (11 °F), with a coverage factor of $k = 2$.

Table 2-1. Number and types of RFID tags examined in the bench-scale experiments; the types of tags are listed in Table 1-1

| Type   | PA1 | PA3 | PA4 | AW1 | AW2 |
|--------|-----|-----|-----|-----|-----|
| Number | 3   | 2   | 2   | 4   | 4   |

# 3 Large-Scale Fire Tests

The bench-scale radiant panel experiments provided a baseline for tag performance when subjected to heating, while the large-scale experiments provided transient heating and combustion product exposure more similar to conditions encountered during firefighting.

## 3.1 Test Specimens

Large-scale fire tests were conducted using both active and passive RFID tags. All passive tags were Gen 2 and operated on the 900 MHz radio frequency range. As shown in Figure 3-1 and Figure 3-2, a variety of tags were used. The number of each type of tag tested in these experiments is shown in Table 3-1. The plastic tags labeled PV1, showing the NIST/BFRL logo were also Gen 2 passive tags. The tags labeled PD1, shown next to the NIST/BFRL, tags were passive high temperature tags and the tags labeled PI1, shown on the bottom of Figure 3-2 were passive specialty tags for use on metal containers. Examples of active tags are shown at the top of Figure 3-2. Two tags labeled AW1 are personal identification tags and one labeled AW2, exhibiting the wire antenna, is a vehicle tag. Both of the active tags operated in the 400 MHz radio frequency range.

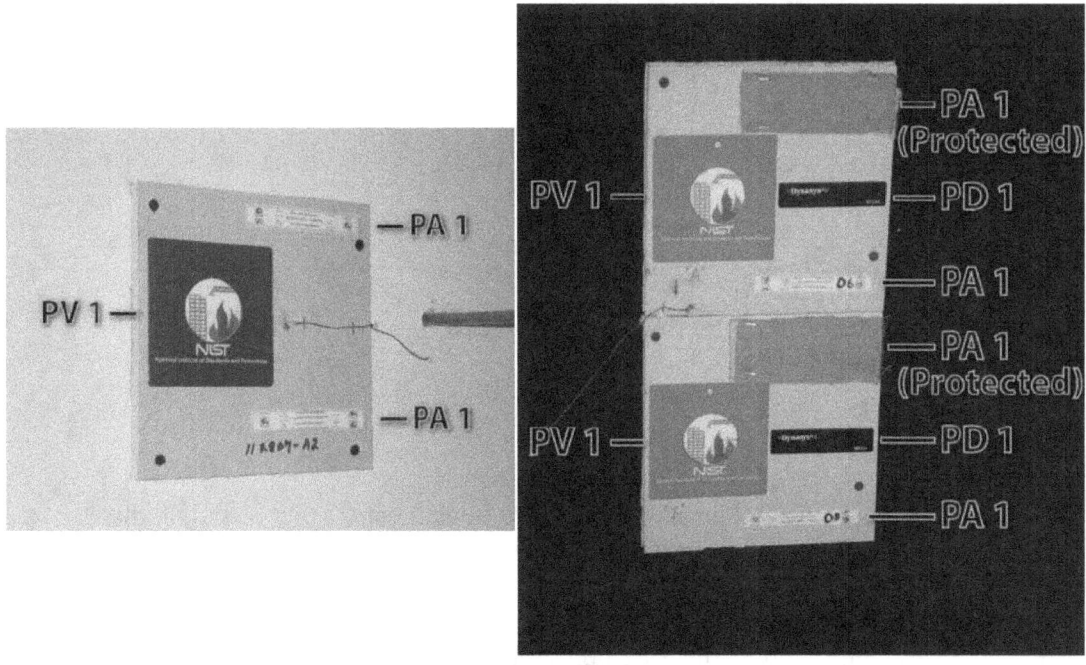

Figure 3-1. Photographs show examples of Passive RFID tags before fire testing.

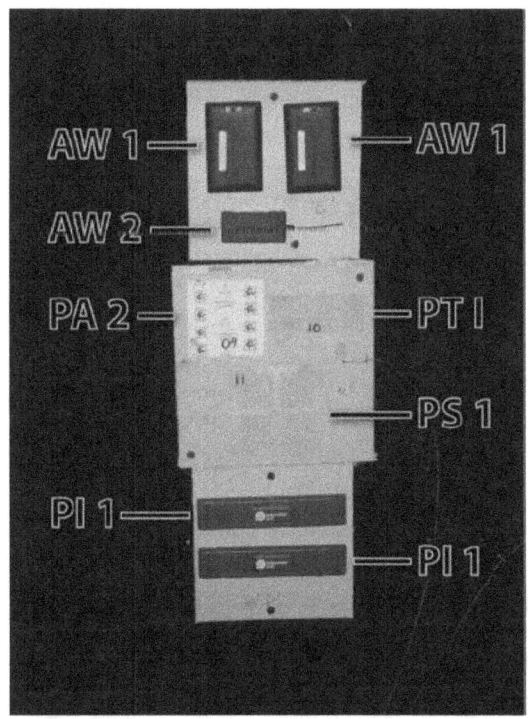

Figure 3-2. Active (top) and Passive (center and bottom) RFID tags before fire testing.

Table 3-1. Number of RFID tags examined in each large-scale experiment arranged by type of tag; the types of tags are listed in Table 1-1

|  | PA1 | PA2 | PI1 | PD1 | PS1 | PTI | PV1 | AW1 | AW2 |
|---|---|---|---|---|---|---|---|---|---|
| Experiment 1 | 10 | | | | | | 5 | | |
| Experiment 2 | 10* | | | | | | 5 | | |
| Experiment 3 | 2* | | | | | | 5 | 1 | 1 |
| Experiment 4 | 4* | 1 | 2 | 2 | 1 | 1 | 2 | 2 | 1 |
| Experiment 5 | | | | | | | | 4 | 2 |

*In these experiments half of the PA1 RFID tags were protected by a swatch of PPE material

## 3.2 Test Instrumentation

RFID tag temperatures for all tests were measured using bare, Type K, 0.254 mm (0.010 in) nominal diameter wire thermocouples. These thermocouples were attached to gypsum board specimen mounting boards, which served as a base for the RFID tags. Temperatures reported for RFID tag thermal performance are estimates based on the specimen board thermocouple measurements. The thermocouples used for these measurements were typically located at least 50 mm away from the RFID tags in an attempt to prevent them from influencing the reception and transmission of radio communications between the RFID tag and the reader. Distance from the RFID tag to the thermocouple ranged from 15 mm (0.6 in) to 305 mm (12 in). Examples of thermocouple attachment points can be seen in several of the figures located in this report (for example, Figure 3-1, Figure 3-2, and Figure 3-5). Temperatures in this report represent estimated surface temperatures for a given RFID tag.

### 3.2.1 Passive Tag Instrumentation

Passive RFID tags were read using two different systems. The first system consisted of a computer with RFID reader software, a 9 volt battery powered transmit/receive control card mounted in a plastic utility box, and an antenna that measured approximately 146 mm by 146 mm (5.7 in by 5.7 in) square. This system is shown in Figure 2-1. This small system, producing < 25 milliwatts output power was found to be inadequate for the large-scale test apparatus; a larger more powerful system was used for these tests. The second reader system, with an output power up to 2 watts, consisted of a computer with RFID reader software, a 120 volt AC power supply, a transmit/receive control box, and a much larger antenna that measured approximately 711 mm (28 in) high and 305 mm (12 in) wide. (Figure 3-3)

In Figure 3-3, the passive reader transmit/receive control box is located on the table behind the computers and was connected to the computer by wire cables. This larger system worked well when reading tags from a distance of about 5 m (16 ft). The maximum range for this larger reader system when reading passive tags through a nominally 38 mm (1.5 in) thick gypsum board wall was about 8 m (26 ft). Both of the RFID reader systems described above interrogated the RFID tags several times a second.

### 3.2.2 Active Tag Instrumentation

The active tag reader consisted of a computer with RFID reader software and a 430 MHz band radio receiver. The receiver had an integrated 80 mm (3.4 in) long antenna built into the unit. (Figure 3-3) This system could receive a 300 microwatt RFID signal through a nominally 38 mm (1.5 in) thick gypsum board wall at more than 10 m (33 ft). The computer and RFID receiver were connected by a wireless "Bluetooth" data link.

### 3.2.3 Measurement Uncertainty

There are different components of uncertainty in the length, temperature, heat flux, gas concentration, differential pressure, gas velocity, and heat release rate reported in this report. Uncertainties are grouped into two categories according to the method used to estimate them. Type A uncertainties are those which are evaluated by statistical methods, and Type B are those which are evaluated by other means.[8] Type B analysis of systematic uncertainties involves estimating the upper (+ a) and lower (- a) limits for the quantity in question such that the probability that the value would be in the interval (± a) is high. After determining uncertainties by either Type A or B analysis, the individual uncertainties are combined in quadrature to yield the combined standard uncertainty. Multiplying the combined standard uncertainty by a coverage factor of two results in the expanded uncertainty, which is taken to correspond to a level of confidence of approximately 95 %. For some of these components, such as the zero and calibration elements, uncertainties are derived from instrument specifications. For other components, such as differential pressure, past experience with the instruments provided input in the uncertainty determination.

All length measurements were taken carefully. Length measurements such as the room dimensions, instrumentation array locations and fan placement were made with a hand held laser measurement device which has an accuracy of 6.0 mm (0.25 in) over a range of 0.61 m (2.00 ft) to 15.3 m (50.0 ft).[9] However, conditions affecting the measurement, such as levelness of the device, yield an estimated expanded uncertainty of 0.5 % for measurements in the 2.0 m (6.6 ft) to 10.0 m (32.8 ft) range. Steel measuring tapes with a resolution of 0.5 mm (0.02 in) were used to locate individual sensors within a measurement array and to measure and position the furniture. Some issues, such as "soft" edges on the upholstered furniture, resulted in an estimated expanded uncertainty of 1.0 %.

---

[8] Taylor, B.N., and Kuyatt, C.E., "Guidelines for Evaluating and Expressing the Uncertainty of NIST Measurement Results", National Institute of Standards and Technology, Gaithersburg. MD., NIST TN 1297, January 1993.
[9] Stanley Hand Tools, User Manual TLM 100, 1000 Stanley Drive, New Britain, CT 06053.

The standard uncertainty in temperature of the thermocouple wire itself was 2.2 °C at 277 °C and increased to 9.5 °C at 871 °C as determined by the wire manufacturer.[10] The variation of the temperature in the environment surrounding the thermocouple is known to be much greater than that of the wire uncertainty.[11,12] Small diameter thermocouples were used for all temperature measurements to limit the impact of radiative heating and cooling. The estimated expanded uncertainty for temperature in these experiments was 15 %.

In this study, total heat flux measurements were made with water-cooled Schmidt-Boelter gauges. The manufacturer reports a 3 % calibration expanded uncertainty for these devices.[13] Results from an international study on total heat flux gauge calibration and response demonstrated that the expanded uncertainty of a Schmidt-Boelter gauge is typically 8 %.[14]

In the following sections, the measurements will be presented in graphic and tabular form. In the graphs, an error bar represents the estimated expanded uncertainty of the measurement. In the tables, the uncertainty is included as part of the caption.

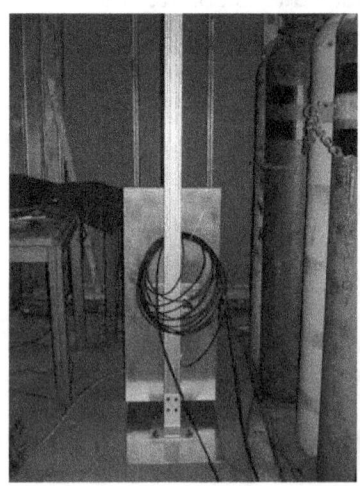
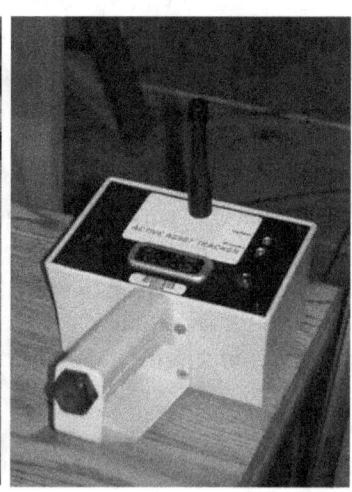

---

[10] Omega Engineering Inc., The Temperature Handbook, Vol. MM, pages Z-39-40, Stamford, CT., 2004.
[11] Blevins, L.G., "Behavior of Bare and Aspirated Thermocouples in Compartment Fires", National Heat Transfer Conference, 33rd Proceedings. HTD99-280. August 15-17, 1999, Albuquerque, NM, 1999.
[12] Pitts, W.M., Braun, E., Peacock, R.D., Mitler, H.E., Johnsson, E.L., Reneke, P.A., and Blevins, L.G., "Temperature Uncertainties for Bare-Bead and Aspirated Thermocouple Measurements in Fire Environments," Thermal Measurements: The Foundation of Fire Standards. American Society for Testing and Materials (ASTM). Proceedings. ASTM STP 1427. December 3, 2001, Dallas, TX.
[13] Medtherm Corporation Bulletin 118, "64 Series Heat Flux Transducers," Medtherm Corporation, Huntsville, AL. August 2003.
[14] National Type Evaluation Program, Certificate Number 00-075A1, Model K-series, Mettler-Toledo, Worthington, OH., December 27, 2002

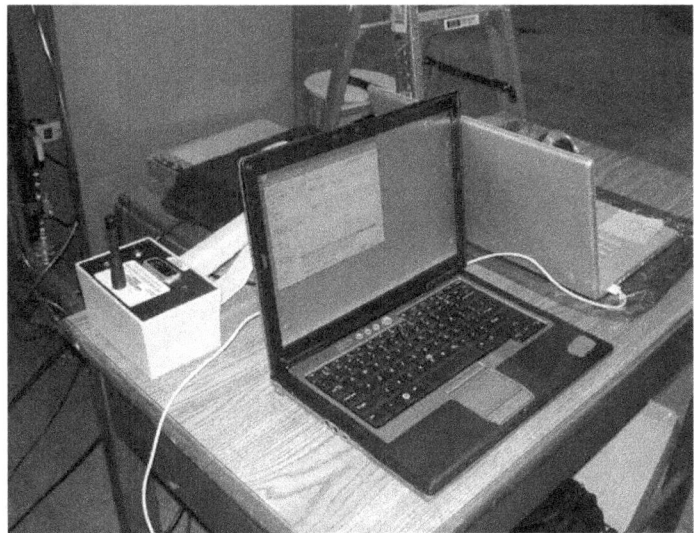

**Figure 3-3. Photographs showing passive antenna (top left) active antenna and receiver (top right) and computers used for data logging (bottom).**

## 3.3 Descriptions of Large-Scale Fire Tests

Five large-scale structural fire tests were conducted to evaluate the thermal performance of RFID tags at the NIST Large Fire Laboratory[15]. These tests were run in conjunction with a series of tests conducted for evaluating "Wind Driven Fire" behavior inside an apartment style structure. (Figure 3-4) The multi-room facility nominally measured 12.0 m (36 ft) long and 4.8 m (16 ft) wide. The ceiling height was 2.4 m (8 ft) high, and the corridors were 1.2 m (4 ft) wide. Each of the rooms located on opposite ends of the center corridor were 3.7 m (12 ft) wide by 4.8 m (16 ft) long and the center corridor measured 3.8 m (12.5 ft) long. Fires were always started in the bedroom (east room) with a wastepaper basket located next to the bed. The fire was typically allowed to grow until the bedroom window broke, and air was forced at different test velocities through the window by a large fan. Flames traveled down the center corridor and into the living room (east room) where furnishing became ignited. Flames exited the living room door and entered the cross corridor where heat and flame exhausted through the ceiling vent shown in the figure. However, heat and flame did enter the south end of the corridor during some tests causing significant damage to the interior finish and materials being evaluated. Figure 3-5 shows a series of photographs illustrating the rooms before the fire test, during the burning phase, and conditions at the end of the test.

The photographs in Figure 3-5 show the fire tests were usually very intense with significant fire damage occurring even in the east end cross corridor.

---

[15] The fire test series, which provided the fire environment for testing the RFID tags, is described in detail in: Madrzykowski, D, and Kerber, S. "Fire Fighting Tactics under Wind Driven Conditions: Laboratory Experiments," National Institute of Standards and Technology, NIST TN 1618, January 2009.

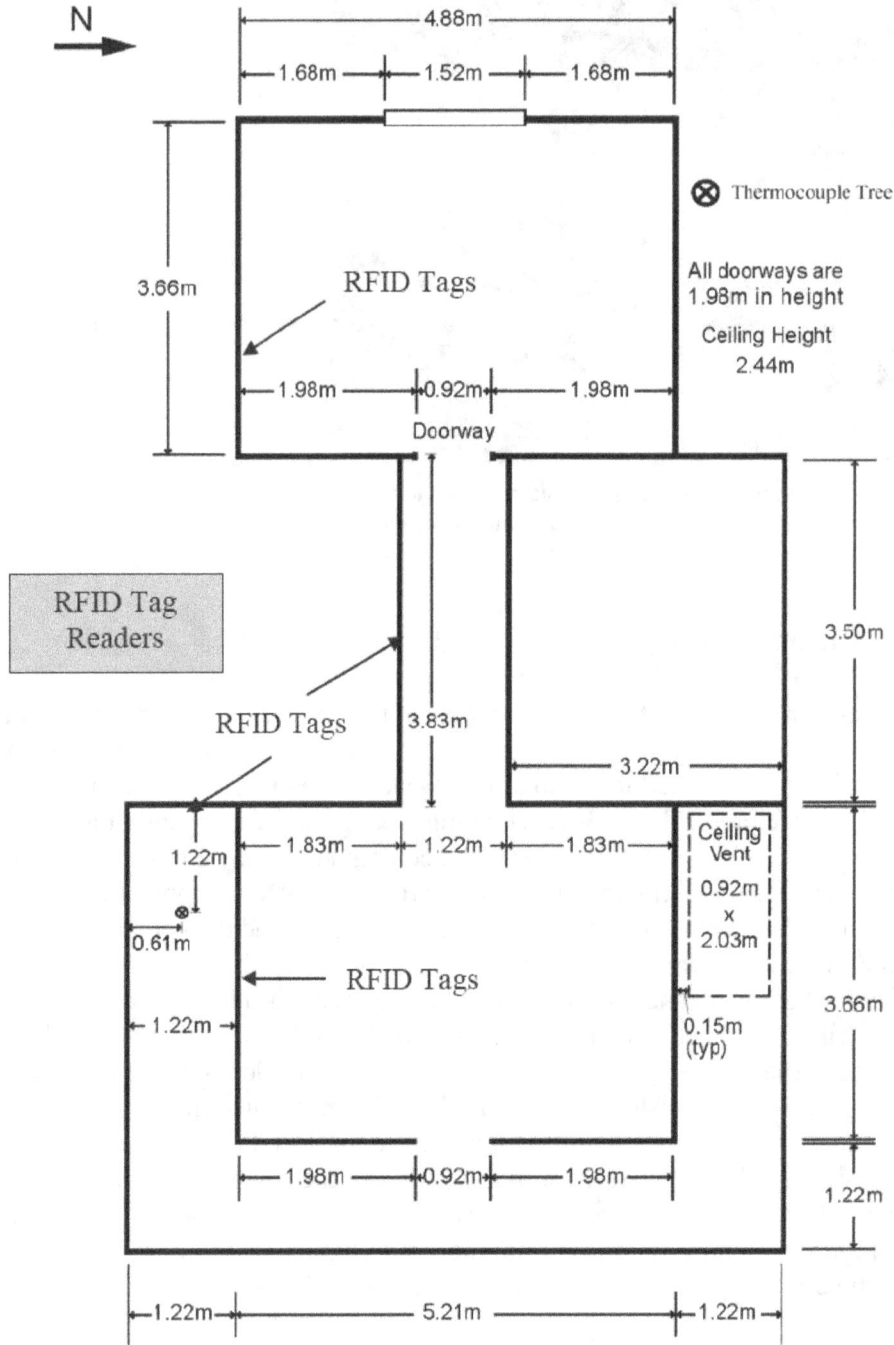

Figure 3-4. Diagram of fire test facility layout and final location for testing RFID tags. The diagram shows the bedroom (top), living room (bottom), and center hallway connecting them. Note: Sketch is not to scale.

Figure 3-5. Photographs show the bedroom (top left), the living room (top right), the center hallway (center left) before the test; the fire in the bedroom (center right), the bedroom (lower left) and the end corridor (lower right) after the fire test. Tags PA 1 and PV 1 are shown in Figure 3-1 and Figure 3-2.

### 3.3.1 Initial Fire Tests #1 and #2

The initial fire tests were used as a means to determine the survivability of RFID tags in spaces that may become fully involved in fire or have very high heating levels. In this case, two different types of Gen 2 passive RFID tags were attached to 203 mm x 203 mm x 13 mm (8 in x 8 in x ½ in) pieces of gypsum board. These specimen boards were then screw attached to the test facility walls. The photograph on the left side of Figure 3-1 shows the initial basic test arrangement. As can be seen, there were two Gen 2 plastic film tags and one NIST/BFRL logo Gen 2 plastic tag. In addition, the figure shows the thermocouple attached to the center of the specimen board. This thermocouple was used to determine the approximate exposure temperature experienced by the RFID tag specimens.

During the initial fire tests passive RFID tags were placed on the lower part of the walls on the inside of the bedroom opposite the bed, in the center hallway, and in the living room. These three walls were located on the south side of the test structure. Examples of RFID tag test specimen placement are shown in Figure 3-1 and the sequence photographs of Figure 3-5. Placement of the RFID specimens in the bedroom and living room were at two locations each: 0.9 m (3 ft) from the floor, and directly above that location 1.8 m (6 ft) from the floor. The center hallway had one specimen board located in the center of the hallway 0.9 m (3 ft) above the floor, and the cross corridor at the east end of the test structure had one specimen board located at the center of the adjacent living room doorway 0.9 m (3 ft) above the floor.

After experiment 1, the Gen 2 passive tags were insulated from thermal exposure. The photograph on the right side of Figure 3-6 shows a passive tag covered with a two layer covering made from fire fighters' protective clothing fabrics. These protective coverings consisted of an outer layer of a common fire fighters' protective clothing shell fabric which covered a quilted fabric thermal barrier with the RFID tag located under it and attached by adhesive tape to the gypsum wallboard specimen board.

### 3.3.2 Fire Test #3

Five of the square, thermo-plastic encapsulated, Gen 2 passive RFID tags were evaluated during this experiment. The tags were molded to the inside of a thermo-plastic material that exhibited the NIST/BFRL logo on its surface. Three of the tags were blue plastic and two tags were red plastic. There was no difference in these tags except for color. See examples in Figure 3-1. In addition, two plastic film backed stick-on Gen 2 tags were tested. One of the plastic film stick-on tags was covered with a two layer covering of fire fighters' protective clothing fabrics. Four of the NIST/BFRL logo thermo-plastic encapsulated tags were located approximately 0.9 m (3 ft) above the floor, and one was located about 0.6 m (2 ft) above the floor. Additionally, two active RFID tags were evaluated and were located approximately 0.76 m (2.5 ft) above the floor. One of the active tags was a personal identification tag, and one was a vehicle style tag with a visible wire antenna. See Figure 1-1 and Figure 3-2.

### 3.3.3 Fire Test #4

Thirteen Gen 2 passive RFID tags were evaluated during fire test #4. Seven of these tags were plastic film based, two were the square NIST/BFRL logo encapsulated plastic tags, two were rectangular blue plastic encapsulated metal container tags, and two were black plastic encapsulated high temperature tags. Two of the plastic film based stick-on tags were covered with a two layer covering of fire fighters' protective clothing fabrics. This test also included three active RFID tags, two personal identification tags and one vehicle tag with a wire antenna. All tags were located on the wall between 0.6 m (2 ft) and 0.9 m (3 ft) above the floor. See Figure 3-13.

### 3.3.4 Fire Test #5

No Gen 2 passive RFID tags were evaluated during test #5. Six active RFID tags were evaluated, four personal identification tags and two vehicle tags with wire antennas. These tags were located 0.7 m (2 ft 3 in) above the test facility floor level.

## 3.4 Test Results and Discussion

### 3.4.1 Results of Initial Fire Tests #1 and #2

Results from the first two RFID tag fire tests showed that the passive tags could not withstand intense fire and thermal conditions when exposed directly to a structural fire environment. Temperatures in Tests 1 and 2, where the RFID tags were located exceeded 600 °C ±16 °C (1112 °F ±29 °F). All RFID tags were completely destroyed by the fire. Figure 3-6 shows one of the test specimen boards in the left photograph following one of these fire tests. It shows that the RFID tags were completely burned away. However, in the photograph to the right, it is evident that the lower RFID tag was destroyed, but the one protected by the two layers of fire fighters' protective clothing fabrics remained in place. Additionally, in the lower right photograph of Figure 3-5 showing the cross corridor, a small white spot can be seen on the right wall across from the doorway. This white spot shows where a RFID tag specimen board was located. The board was so badly damaged that all RFID tags were completely destroyed and the board fell from the wall after the test. Temperatures during these thermal exposures exceeded 200 °C ±16 °C (392 °F ±29 °F). No recordable electronic data on RFID tag performance are available from these early fire tests as a result of the RFID reader's inability to read the widely dispersed tags in the test structure. However, the passive RFID tags which were located under the protective clothing fabric often worked after being removed from the test facility.

 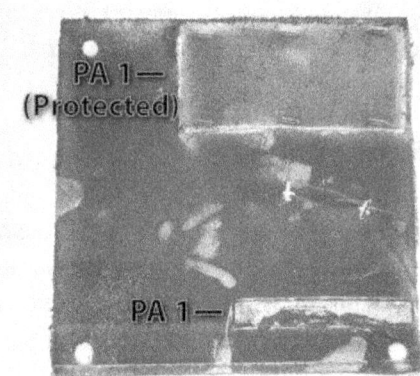

**Figure 3-6. Photographs showing passive RFID tag after fire exposure, the tags on the left specimen board had burned away.**

It was learned from the initial tests that the RFID antenna system and reader were not adequate to obtain consistent readings from the passive tags. The same antenna and reader system shown in Figure 2-1 was used for the early tests, and it lacked the performance for measuring the RFID tag signal behavior. In the later tests, a larger more powerful reader and antenna system was applied to the experiments, and the RFID tags were moved to a location in the test structure that allowed for better signal performance measurements.

### 3.4.2 Results of Final Large-scale Fire Tests

The last three fire tests of the RFID series, Tests 3, 4, and 5, provided the most reliable data. Both passive and active RFID tags were exposed to thermal conditions in the dead-end corridor which was located on the southeast side of the test structure, see Figure 3-4. The tags were located a minimum of 0.6 m (2 ft) above the test facility floor. The RFID tag experimental location for these fire tests is shown in Figure 3-4. Since the initial tests showed that RFID tags did not survive flame impingement, this location provided some protection from the flame environment and, like in much of the remaining structure, was out of the flame's vent path. The results of these experiments are summarized in Table 3-2. Test data from this location showed that extensive heating occurred, and that smoke was deposited in a heavy layer on all surfaces in the area. Fire test temperature conditions in the dead-end corridor are shown in Figure 3-7, Figure 3-9, and Figure 3-11. These temperatures were obtained from a thermocouple tree extending from the ceiling to the floor and located in the center of the dead-end corridor. Temperature measurement points for this thermocouple tree are shown on the data plot legends, starting at the ceiling and extending towards the floor. Nomenclature for this legend includes the color line for each thermocouple that was located in the center of corridor followed by the measured distance from the ceiling, in meters below the ceiling. Tag exposure temperatures are provided on the accompanying test series plots, and locations for these tags are given in the above test descriptions.

Table 3-2. Results of large-scale tests 3 through 5; $T_{max}$ indicates the maximum temperature recorded at 0.6 m above the floor in the southwest corridor. All times indicate time after the beginning of the test. The estimated expanded uncertainty for temperature in this table is 15 %. For times given, the estimated error is ± 5 s.

|  | Active Tags | Passive Tags | $T_{max}$ |
|---|---|---|---|
| Test 3 | All tags functioned sporadically after 160 s. | All tags stopped functioning between 280 s and 310 s, function did not return during the test. | 330 °C |
| Test 4 | All tags functioned throughout the experiment. | One tag sporadically failed, but this behavior was independent of test conditions. All other tags functioned throughout the experiment. | 260 °C |
| Test 5 | Four of the tags failed during the experiment. Failures occurred between 300 s and 450 s. Two tags operated throughout the experiment. | None used | 360 °C |

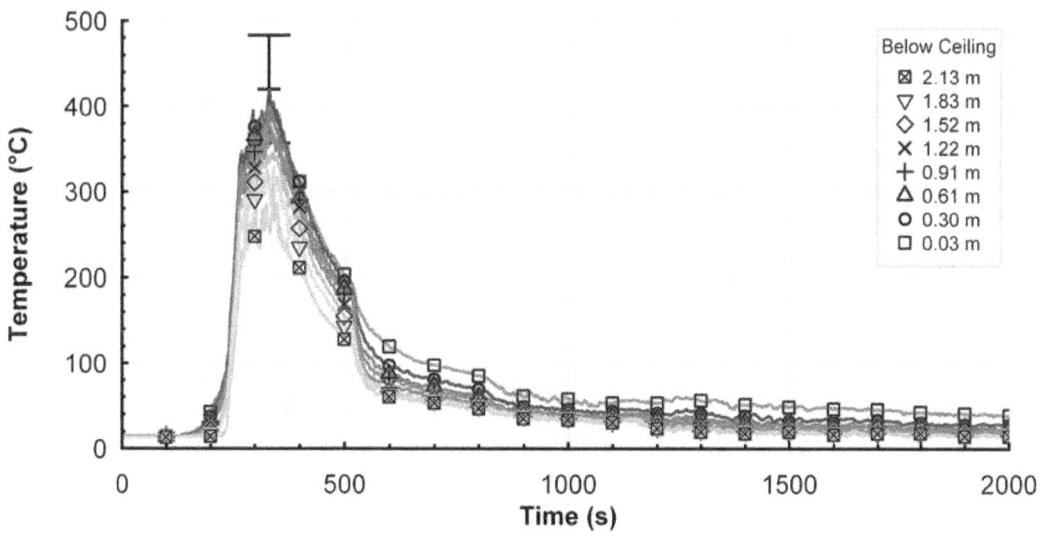

Figure 3-7. Test 3, temperatures center of south corridor housing RFID tags.

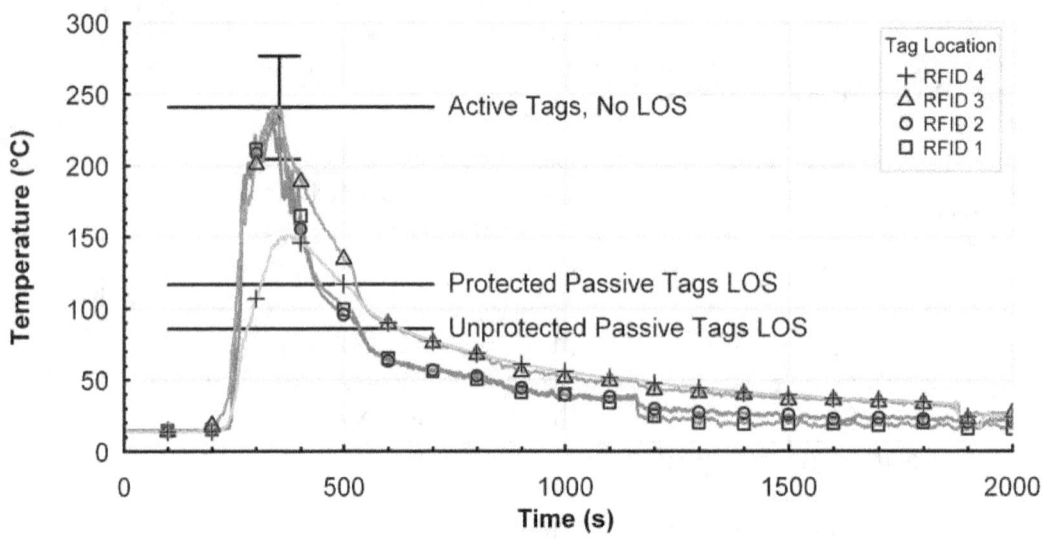

Figure 3-8. Test 3, temperatures on wall next to RFID tag specimens. Loss of signal (LOS) was observed for passive tags; LOS was not observed for active tags.

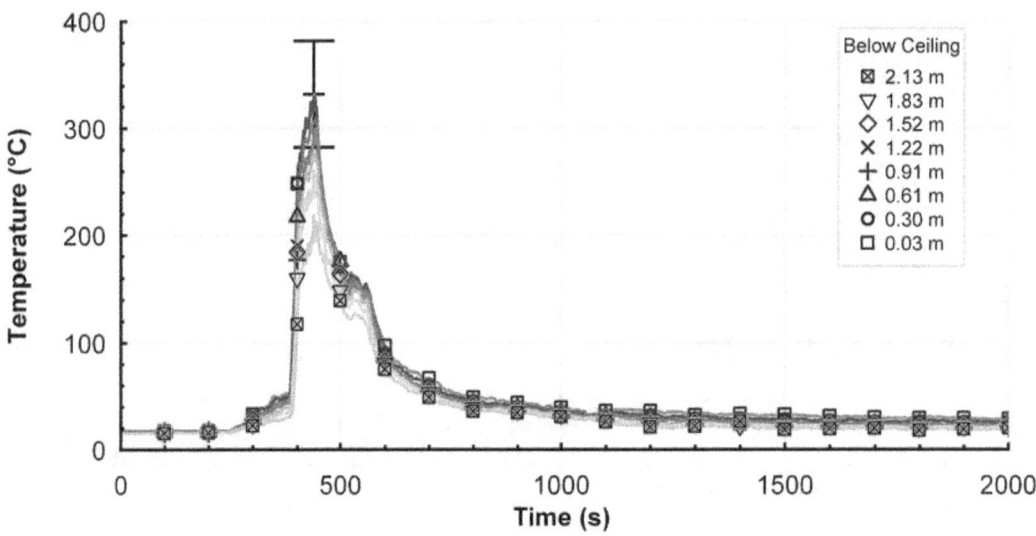

Figure 3-9. Test 4, temperatures center of south corridor housing RFID tags.

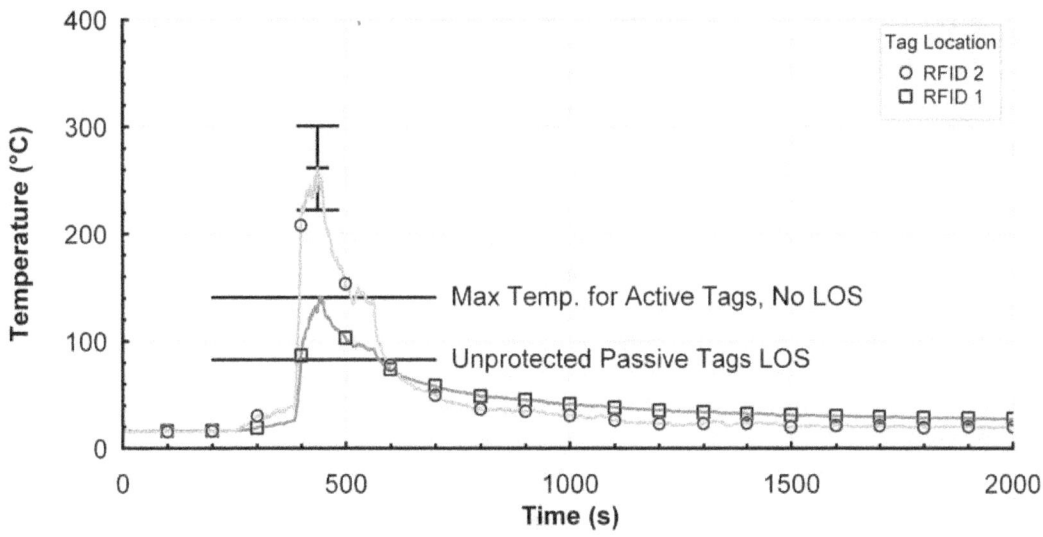

Figure 3-10. Test 4, temperatures on wall next to RFID tag specimens. Loss of signal (LOS) was observed for passive tags.

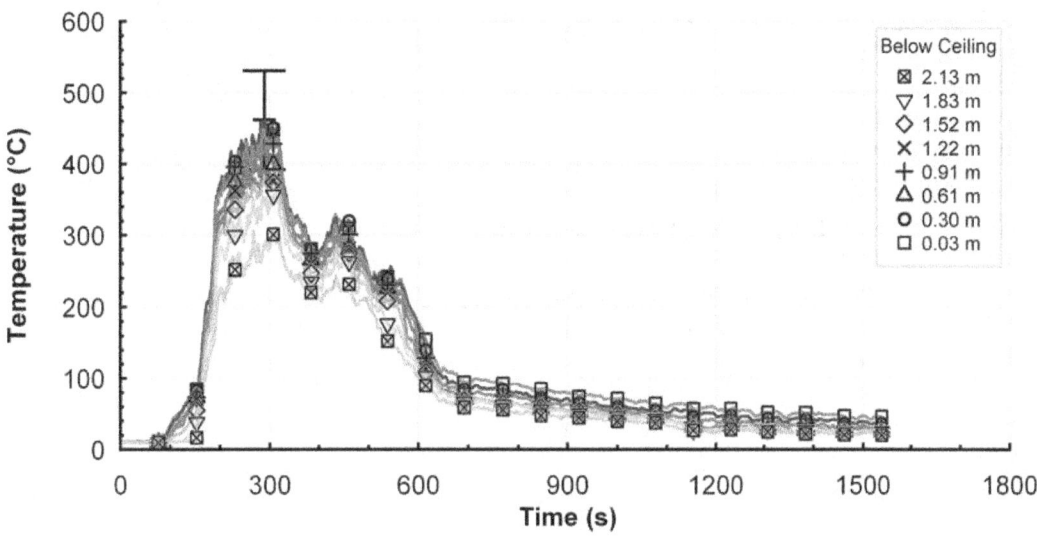

Figure 3-11. Test 5, temperatures center of south corridor housing RFID tags.

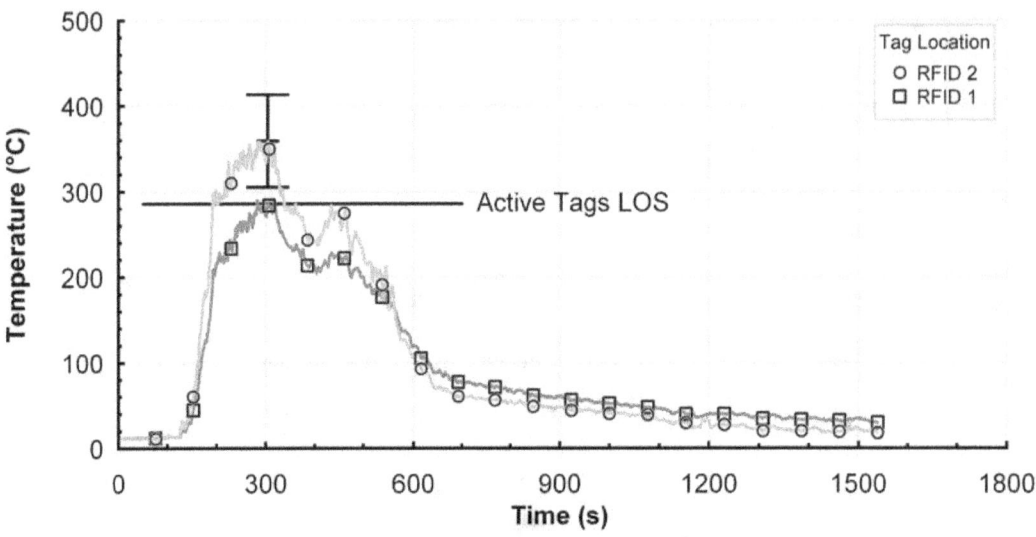

Figure 3-12. Test 5, temperatures on wall next to RFID tag specimens. No passive tags were tested; loss of signal (LOS) was observed for the active tags.

Additionally, temperature measurements at the RFID tag locations on the wall are shown in Figure 3-8, Figure 3-10, and Figure 3-12. In these figures, the thermocouple tree and RFID tag test location temperatures for each test are paired for easy comparison. The legend for these data plots show the color symbol for each of the RFID tag measurement location thermocouples, and it identifies the thermocouple and each specimen board to which the thermocouple is attached. In Figure 3-8, Figure 3-10, and Figure 3-12, it is noted that thermocouple temperatures vary and that some are lower than others. This results from the relative locations of the specimen boards and thermocouples to the hot gas flow in the dead-end corridor. Estimated RFID tag performance temperatures reported in this text are drawn from the thermocouple located closest to the given RFID tag.

### 3.4.2.1 Results RFID tag Test #3

*Passive Tags*

In RFID tag test #3, the unprotected passive RFID tags failed at 86 °C ±16 °C (187 °F ±29 °F) and the protected RFID tags NIST/BFRL encapsulated in plastic or covered with fire fighter's protective clothing materials began to fail at temperatures of about 117 °C ±16°C (242 °F ±29 °F ). The NIST/BFRL plastic encapsulated tags melted and fell on the test facility floor. At the end of the test when the tags had cooled back to room temperature, the charred and fragile tags were carefully removed from the floor and placed back against the wall where they had been mounted. Attempts to read the charred tags were successful. As with the earlier small scale laboratory thermal testing, the tags were readable again after cooling back to room temperature. The unprotected plastic film tag remained in place and one end curled away from the attachment point, and it also began working again after cooling down.

### Active Tags

The active RFID tags were readable throughout the test, but as the fire temperatures increased each of the tags under study experienced a signal dropout that lasted approximately five to ten seconds. Each of the signals returned and remained sporadic throughout the remainder of the test. The active RFID tags continued to work while test location temperatures exceeded 220 °C ±16 °C (428 °F ±29 °F). At the end of the test, the active RFID tags showed significant physical thermal damage but returned to normal function after cooling to room temperature. Examples of the types of thermal damage are shown in Figure 3-13. The photographs exhibit melting and bubbling of the plastic cases.

#### 3.4.2.2  Results RFID tag Test #4

### Passive Tags

All passive RFID tags worked throughout test #4, except one. This tag failed to respond at approximately six and one-half minutes into the test when the test location temperature was approximately 83 °C ±16 °C (181 °F ±29 °F). The tag began to respond again in a sporadic fashion about eight minutes later when the test location temperature dropped below 37 °C ±16 °C (99 °F ±29 °F). Results from this test were surprising since none of the passive tags showed any significant thermal damage and thermocouple temperatures at the test location had a brief peak that exceeded 250 °C ±16 °C (482 °F ±29 °F). It is postulated that the peak temperature duration was too short to cause thermal damage to the tags that were attached to the gypsum board substrate which acted as a heat sink.

### Active Tags

All three active RFID tags survived the test and provided steady readings throughout the thermal exposure. The two personal tags showed thermal deformation, and the single vehicle tag with the wire antenna showed no visible thermal damage. The tags experienced temperatures that peaked at approximately 141 °C ±16 °C (286 °F ±29 °F). All tags were covered with a thick layer of soot from smoke at the test location.

#### 3.4.2.3  Results RFID Tag Test #5

### Passive Tags

No passive tags were evaluated during this experiment.

### Active Tags

Four of the six active tags stopped working during the fire test exposure. Both wire antenna tags melted away and fell from the specimen boards. All of the personal tags were deformed by heat but remained on the specimen boards. At the end of the test only the two tags located on the lower half of the specimen boards continued to work. The four tags that failed during the test stopped working when the test location temperature exceeded 286 °C ±16 °C (547 °F ±29 °F).

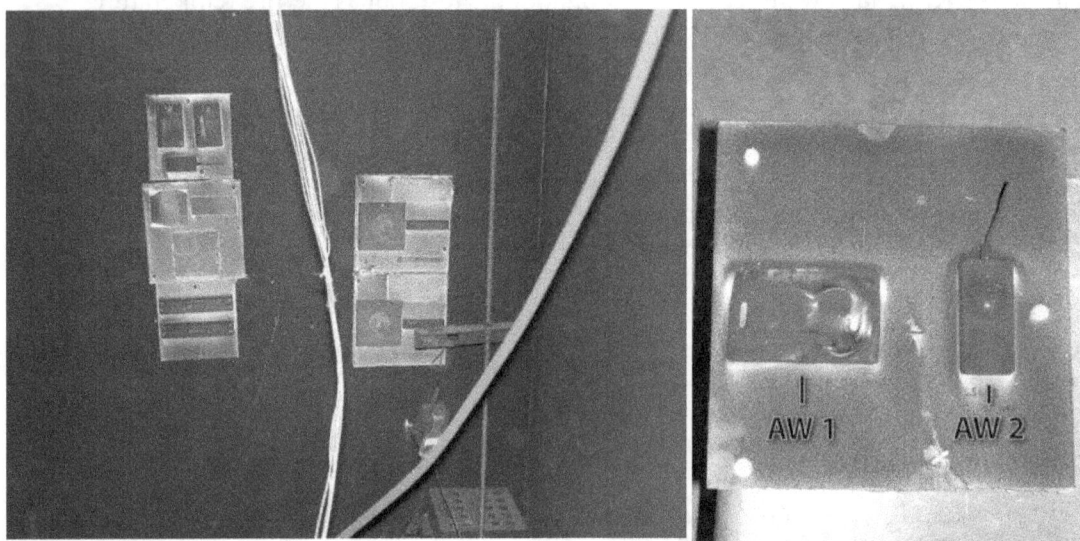

Figure 3-13. Photographs showing fire test damage to passive and active RFID tags. Note: RFID tag identification for these photos is found in Figure 3-1 and Figure 3-2.

# 4 RFID Tag, Wet Experiments

Three different types of RFID tags were examined in the RFID tag wet experiments. The number of each type of tag that was used is listed in Table 4-1.

Table 4-1. Number and types of RFID tags examined in the RFID tag wet experiments; the types of tags are listed in Table 1-1

| Type | PA1 | PV1 | AW1 |
|---|---|---|---|
| Number | 2 | 1 | 1 |

## 4.1 Passive Tag Systems

### 4.1.1 RFID Equipment Used

Three different Gen 2 passive tags were used. One RFID tag was sealed in plastic, AW1, and two tags were RFID circuits placed on film with one sticky side for attachment, PV1 and PA1. The passive tag reader was a Mercury5 system using the large pedestal stand antenna. See Figure 4-1. This is the same system used in the large-scale fire experiments. Dimensions for this antenna are given in section 3.2.1. Figure 4-1 shows the passive tag reader antenna with an aluminum meter stick standing vertically along the left side. Communications wire ran from this antenna to the reader control box and then to a laptop computer.

**Figure 4-1. Passive RFID tag reader antenna used for wet fabric study.**

### 4.1.2 Experiments

After the tag reader was configured and running, the two sticky film based Gen 2 tags, weighing approximately 0.3 g each, were attached to a piece of fabric that was cut from a white cotton knit T-shirt. See Figure 4-2. The T-shirt fabric specimen without the RFID tags attached weighted about 3.4 g when dry and measured approximately 120 mm by 120 mm (4.7 in by 4.7 in) square. The T-shirt material was folded so that only one layer of knit fabric covered the passive tag's surface, see Figure 4-2 below. First, two film backed tags wrapped in dry fabric were placed in front of the antenna, and the reader immediately recorded readings from the tags. The tags were slowly moved away from the antenna, and the reader continued to read the tags until they were at a distance of approximately 12 m (40 ft) from the antenna. At this distance the passive tag readings became sporadic and stopped reading after passing the 12 m (40 ft) mark. Following development of this baseline data, with the tags still located inside the T-shirt fabric, they were placed into a container of water thoroughly wetting the fabric. The fabric was allowed to drip until droplets were no longer coming from the fabric. The wet fabric and tag assembly weighed approximately 13.4 g. The tags with the single layer of wet cotton knit fabric covering them were again placed in front of the passive system antenna. At a distance of approximately 152 mm (6 in) the reader only received sporadic signals from the passive tags. Upon moving the tags beyond the 152 mm (6 in) mark the reader failed to receive any signals from the tags.

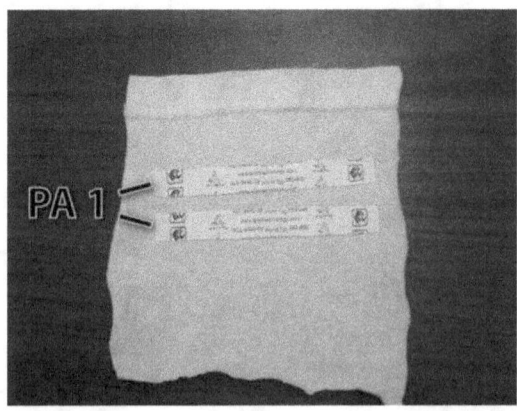 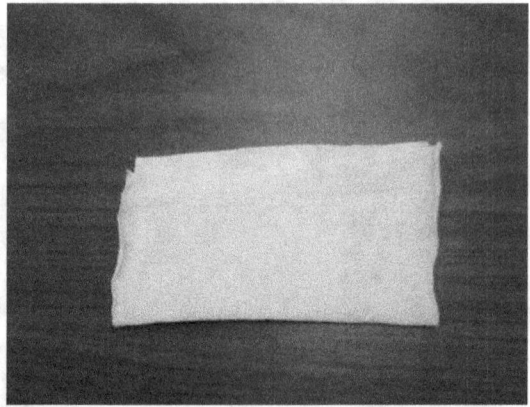

**Figure 4-2. Photographs showing passive (PA1) RFID tags attached to cotton T-shirt fabric (left) and fabric folded over tags (right).**

In order to determine the failure mode associated with reading the wet tags, additional experiments were conducted. Two hypotheses were tested: 1) was the failure to read the tag circuits associated with water located in the fabric and touching the tag physically shorting out the circuit printed on the tag, or 2) did the water in the single layer of wet cotton knit fabric attenuate the radio signal. The single Gen 2 RFID tag sealed in the plastic jacketed, (Figure 4-3) weighing approximately 10.0 g, was tested dry with a single layer of T-shirt material wrapped around it. Since the plastic sealed tag was larger than the film backed tags a larger piece of T-shirt material was needed. This piece of T-shirt material weighed approximately 11.6 g, when dry, and measured approximately 180 mm by 360 mm (7.1 in by 14.2 in). The passive reader system using the large antenna was again able to read the tag up to a distance of about 12 m (40 ft). Following this successful baseline test, the RFID tag, still wrapped in cotton knit fabric was wet by placing it into a container of water. The fabric covered tag was removed from the water and fabric was allowed to stop dripping. The wet fabric and tag assembly weighed approximately 55.8 g. This tag and fabric assembly was placed approximately 150 mm (6 in) in front of the operating passive RFID reader antenna. Again, the tag was only read sporadically. When the tag was moved further away, the reader was unable to detect any signal from the RFID tag. This experiment shows that it was unlikely that the two film backed tag circuits were grounded by the wet fabric and that the failure likely resulted from the water itself causing the RF signal to be attenuated. The plastic sealed tag was removed from the wet T-shirt material and the passive reader was able to read the tag again up to the 12 m (40 ft) mark proving that the tag was still functional. Following this experiment, the two wet film backed tags were tested again, and the reader was unable to read the tags. These tags were then removed from the wet fabric, dried by a cloth, and again the passive reader was able to read the film backed tags up to the 12 m (40 ft) mark showing that the two tags were still functional.

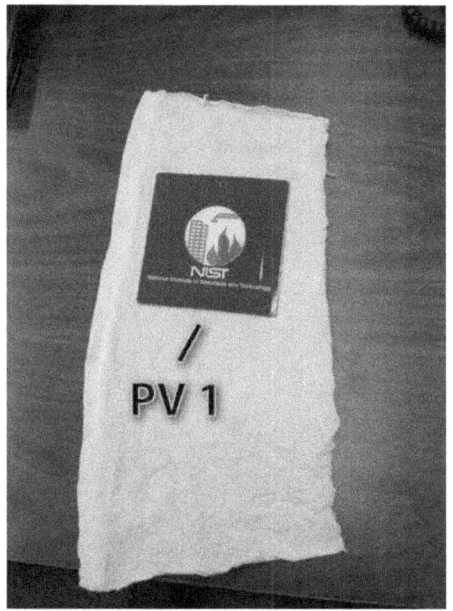 

**Figure 4-3. RFID tag (PV1) encapsulated in plastic with T-shirt fabric (left); RFID tag enclosed in dry T-shirt fabric (right).**

## 4.2 Active Tag Systems

### 4.2.1 RFID Equipment Used

Three active RFID tags were used in the wet and dry tests. Each tag had the electronics enclosed inside a plastic covering. The tags weighed approximately 22 g. See Figure 4-4. A Dynasys Technologies Inc., Wavetrend, Active Asset Tracker with an 89 mm (3.5 in) flexible rubber antenna located on the top can be seen in Figure 4-5. This is the same active tag reader system used in the large-scale fire experiments. The RFID reader was connected to the data computer by a Bluetooth communications link.

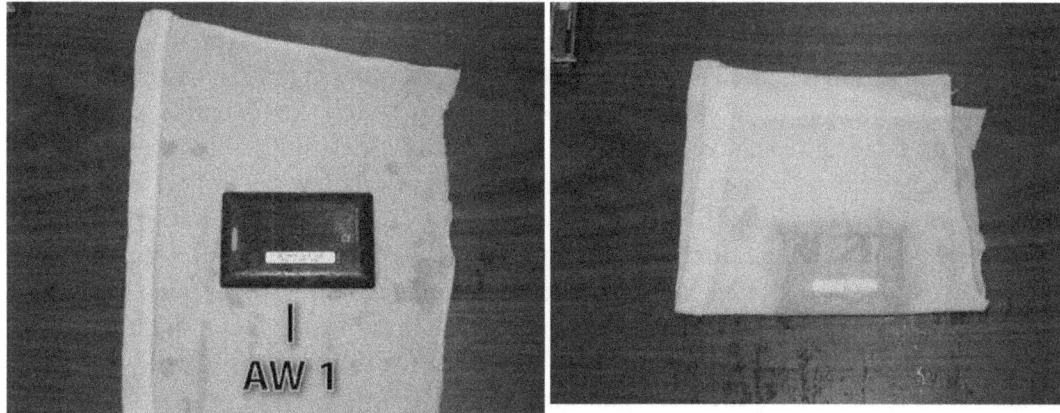

**Figure 4-4. Active RFID tag (AW1) with T-shirt fabric (left) and RFID tag enclosed in wet T-shirt fabric (right).**

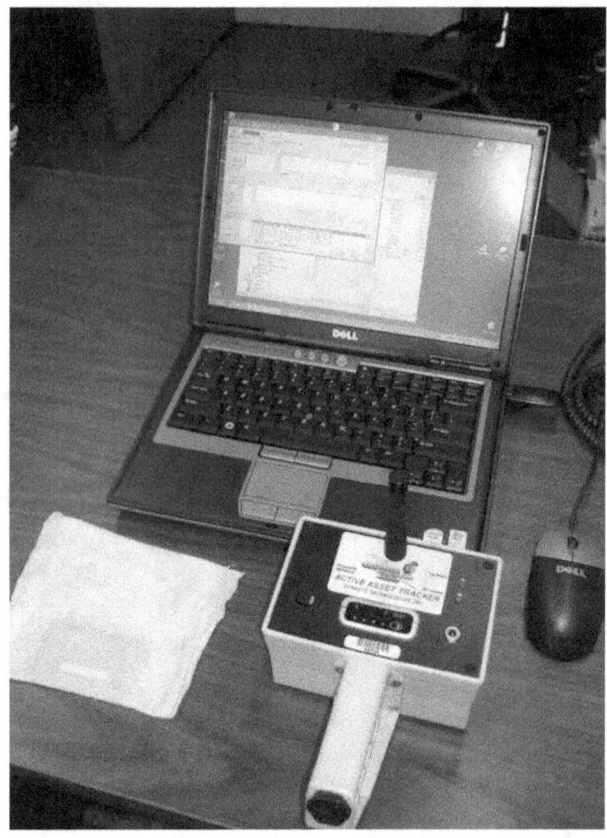

**Figure 4-5. Photograph showing the complete active RFID reader system and one of the wet test specimens.**

### 4.2.2 Experiments

After the active tag reader was configured and operating, a dry active RFID tag was covered by the larger T-shirt fabric described above. This active tag with a single layer of T-shirt knit fabric placed over it was moved away from the reader. The reader was able to read the active tag up to a distance of approximately 17 m (57 ft). As above with the passive tags, the T-shirt fabric was then wetted and allowed to drip until it stopped dripping. The same active tag was wrapped in the wet fabric with a single layer of fabric enveloping the entire tag. The wet fabric contained approximately 45 g of water. Again the tag was moved away from the RFID reader until the tag readings became intermittent. The distance where readings stopped was approximately 14 m (45 ft). The reading difference from dry to wet test conditions was approximately 3 m (10 ft). This indicated that the wet fabric was attenuating the RF signal from the active tag, but it did not prevent it from being read at a significant range as compared to the passive tags which were not readable when covered with the wet T-shirt fabric.

### 4.3 Summary of Wet Experiment Results

The results of the RFID tag wet experiments indicate a set of maximum read distances for the tags tested under wet and dry conditions. Table 4-2 summarizes those results.

**Table 4-2. Maximum read distance for the three tags tested under wet and dry conditions. The estimated expanded uncertainty of distance measurements in this table is 1.0 %.**

| Condition | Passive Types | | Active Types |
|---|---|---|---|
| | PA1 | PV1 | AW1 |
| Wet | 150 mm (6 in) | 150 mm (6 in) | 14 m (45 ft) |
| Dry | 12 m (40 ft) | 12 m (40 ft) | 17 m (57 ft) |

# 5 Discussion and Conclusions

Results from these experiments demonstrated that RFID tags can be completely destroyed by fire exposure. It is also shown that RFID tags will operate within a limited temperature range. Test data show that if an RFID tag stopped working due to an elevated temperature exposure, it may begin functioning again after its temperature returns to the operational range.

**Table 5-1. Recommendations for Thermal Classes of Electronic Equipment Used by First Responders Donelly, et. al.** [16]

| Thermal Class | Max. Time (min) | Max. Temperature (°C / °F) | Max. Heat Flux (kW/m²) |
|---|---|---|---|
| I | 25 | 100/212 | 1 |
| II | 15 | 160/320 | 2 |
| III | 5 | 260/500 | 10 |
| IV | <1 | >260/500 | >10 |

Table 5-1 provides a range of 4 different thermal classes that are recommended for use in evaluating electronic equipment that would be used by fire fighters. A Thermal Class Level III would be typical of a device that a fire fighter would rely on for life safety purposes, for example, a Personal Alert Safety System or "PASS" device. NFPA 1982 requires PASS devices to withstand a High Temperature Functionality Test in order to be certified[17]. This test requires the PASS device to be exposed to a temperature of 260 °C (500 °F) for five minutes in an oven. This same time and temperature benchmark is used here as a point of comparison with the RFID measurements.

The exposures in these experiments were designed to represent real-world, unsteady-heating environments. Rather than being exposed to a relatively consistent temperature,

---

[16] Donnelly, M. K., Davis, W.D., Lawson, J.R., and Selepak, M.J. "Thermal Environment for Electronic Equipment Used by First Responders." National Institute of Standards and Technology, TN 1474 (2006).
[17] *NFPA 1982 Standard on Personal Alert Safety Systems (PASS)*, National Fire Protection Association, Quincy, MA, 2007.

as with oven heating, the tags were exposed to a combination of convective, radiative, and conductive heating and cooling effects. In the small scale experiments, the tags were heated primarily through conduction through contact with the heated substrate surface while being convectively cooled by the ambient air. In the large-scale experiments, the tags were heated through convection and radiation by hot combustion products and cooled through conduction.

Data from the radiant panel tests showed that the limiting temperature where Gen 2 Passive RFID tags failed to function was on the order of 60 °C (140 °F). In the large-scale fire tests, the initial failure temperature appears to be on the order of 80 °C (180 °F). This difference in initial failure temperature between radiant panel and large-scale fire testing is likely associated with differences in transient heating between the radiant panel and large-scale experiments.

The active RFID tags appeared to be more robust, transmitted usable signals a greater distance, and experienced initial thermal exposure failures at temperatures in excess of 200 °C ±16 °C (392 °F ±29 °F). Active RFID reading systems that worked in the large-scale fire test environment were much smaller than the passive reading system that eventually gave satisfactory performance. The antenna and electronic systems for active RFID tags can easily sit on a desk top with a portable computer that operates the reader. As exhibited by the photographs in this report, however, the active RFID tags are typically larger than the plastic film passive tags.

These experiments have shown some of the limitations of RFID tags under potential rough duty environments involving moisture and heat. Signal attenuation due to wet material precludes the passive tags being worn under PPE due to the potential that the fabric may become wet. The thermal failure level of RFID tags is low compared to other equipment standards, such as those for PASS devices. As currently designed and protected, the RFID tags are not able to operate at a temperature of 260 °C (500 °F) for five minutes, and therefore could not be used as a life safety device for a structural fire fighter.

## 5.1 RFID Tags and Emergency Response Operations

The work associated with this research effort has provided insight into the performance of passive and active RFID tags when exposed to environments that may be experienced by emergency responders. Findings from this study provide information on the type of performance that emergency responders should expect from the technology. Also, findings for this generation of RFID tag systems suggest ways that the technology can be applied by the emergency response community. The following is a brief discussion of these issues.

Performance of these two types of RFID tag systems, passive and active, suggests that applications of the devices appear to fall into two different operational domains. However, with advancements in RFID technology, the separation of these two domains may become less apparent. One domain relates to normal day-to-day logistics associated with maintenance and tracking of equipment. The second domain relates to active tracking of personnel and equipment associated with an emergency response. Performance of the two different RFID tag systems, as experienced with this study, tends to segregate the systems for favorable use in either one or the other operational domains. Operational performance, (functional operating range and equipment requirements) when coupled with research findings from thermal and moisture exposures, suggest that the passive RFID tag systems evaluated are more appropriately used for normal day-to-day equipment logistics and tracking. Results from operations with the active RFID tag systems tend to favor their use in the field where emergency response operations require dynamic tracking capabilities.

Findings from these studies also suggest that neither system will be successful without investment in well designed systems and infrastructure. Each system requires tags and reading components. If the system is to be used for personnel tracking in a facility, its structures must be carefully evaluated so that an appropriate system design can be engineered to accomplish the tracking goals. The system designer must determine, for example, whether the tag reader or the tag is the fixed component in the structure. These considerations will drive the design requirements for the remainder of the system, which will likely have a significant impact on the functional capabilities and cost of the system. Additionally, new standards may be needed as the systems mature and emergency responders explore their use. Building codes will need to be updated to insure that the RFID tag systems function properly and safely in all building environments.

## 5.2 Implications for Body Worn RFID Tag Systems

The application of RFID locating technology to body worn systems for emergency responders is a complicated issue. This technology has the potential for determining the presence of an emergency responder at an incident scene and may also be useful for tracking and locating a responder at the scene. The use of passive and active RFID tag systems in this research effort has highlighted a number of points relative to the use of RFID technology for body worn systems. These points are listed below:

<u>Passive Tag Characteristics:</u>

1) Limited to short range operation "less than a meter to a few meters",
2) Small,
3) Light weight,
4) Thin, near two dimensional construction,
5) Flexible,
6) Mechanically fragile,
7) Can be thermally protected,
8) Hardened/rigid tags available,
9) No battery required,
10) Not functional in wet fabric,

11) Large antenna and RF power required to extend usable range beyond a few meters,
12) Significant equipment and logistical field support required due to the relatively large antennas needed for increasing passive RFID system range,
13) Short range capability suggests that emergency responders must carefully plan and mobilize the passive RFID technology in the field, and
14) Passive tags cost less than active tags.

<u>Active Tag Characteristics:</u>

1) Longer range performance than passive tags, "tens of meters,"
2) Small but larger than the simple passive tags,
3) Light weight,
4) Rigid,
5) Mechanically strong,
6) Operates at higher environmental temperatures than passive tags,
7) Battery life 3 to 5 years, requiring performance maintenance plan,
8) Small receiving antenna provides operating reading ranges of tens of meters,
9) Directional antenna available which may be used in triangulation,
10) Marginally functional in wet fabric,
11) Significantly greater operating range, combined with smaller antenna and receiver systems, suggests that active tag system would require less equipment/logistical support than passive systems in the field,
12) Long range performance suggests that less preplanning and mobilization efforts by emergency responders would necessary for successful use of active RFID technology, and
13) Active RFID tags cost more than passive tags.

To insure that emergency responder's safety is not compromised by a failed RFID tag system, each passive and active system will require a test and maintenance programs.

Both data reported in this study, and the comparative lists located above, indicate that active RFID tag systems possess an advantage over passive tag systems for field use. This results primarily from the active tag system having the capability to be operated over a significantly greater range when coupled with small and simple computer based antennas and equipment. Even though the active RFID tags are generally larger than the passive tags, they are still physically small and light. As a result, a body worn active tag is not likely to represent a significant burden to the responder. Additionally, active RFID tags tend to perform at higher temperatures than passive tags.

## 5.3 Recommendations for Future Work

The experiments carried out in this study have helped to characterize the performance of RFID tag systems when exposed to thermal environments that may be experienced by responders during structural fires. It has shown that thermal exposures can degrade the performance of RFID tag systems. Still, there are other fireground conditions that require study to fully quantify the performance of RFID tag systems during emergency response operations. RFID tag systems and other RF systems used in proposed body worn networks should be tested to insure that they will communicate through typical rough-duty conditions found in firefighting environments, i.e. water spray, flames, smoke, and high particulate atmospheres. Performance metrics developed by future work should be evaluated in simulations of real fire ground conditions where personnel movement and fire fighting operations are being conducted. Evaluation and quantification of the performance metrics, in a controlled simulation of fire ground operations, will be a critical element in establishing appropriate standards for the application of RFID tag systems.

# 6 Acknowledgements

This study, conducted by the NIST Engineering Laboratory at Gaithersburg, MD is in cooperation with other studies on RFID technology carried out by the NIST Electromagnetics Division in, Bolder, CO.

Appreciation is extended to Dan Madrzykowski and Steve Kerber for providing the opportunity to conduct the RFID tag experiments in conjunction with their "Wind Driven Fire" experiments. In addition, Madrzykowski provided the test temperature data for the plots shown in this report. Mr. Roy McLane provided invaluable assistance helping to conduct all of the large-scale RFID tag fire experiments. Appreciation is also extended to Matthew Molek, a Student Intern from St. Mary's College of Maryland, who assisted with taking data during the last half of this research effort. Richard Peacock and Adam Barowy, of the NIST Engineering Laboratory, are recognized for their assistance with running the wet passive RFID tag experiments. Appreciation is extended to Michael Love for his insight and comments. Special thanks to Philip Mattson and Bert Coursey of the Department of Homeland Security, Science and Technology Directorate for their support of this project.

www.ingramcontent.com/pod-product-compliance
Lightning Source LLC
Chambersburg PA
CBHW081805170526
45167CB00008B/3337